Trans-Human Generations

The Next Evolution of a Species

by Joshua Free

*An original Systemology Publication
released in the year 2012 by the
Mardukite Truth Seeker Press.*

© 2012, Joshua Free

No part of this publication may be reproduced in any form or by any means, electronic or mechanical, including photocopying, recording, or by any information storage or retrieval system, without permission in writing from the publisher.

Related Books by Joshua Free

SYSTEMOLOGY:
Human, More Than Human (S1)
Systemology: Defragmentation (S2)
Alien Agendas – Anthology (S1+S2)

NECRONOMICON ANTHOLOGIES:
Necronomicon Anunnaki Bible (Year-1)
Gates of the Necronomicon (Year-2)
Necronomicon Workbook (Year-3)

MERLYN STONE ARCHIVES:
Sorcerer's Handbook of Merlyn Stone (M1)
Lost Books of Merlyn Stone (M2)
Wizardry – Anthology (M1+M2)

THE DRUIDISM TRILOGY:
Book of Elven-Faerie (D1)
Draconomicon (D2)
Druidry (D3)
A Complete Course Curriculum in Druidry (D1,2,3)

BEST-SELLING STAND-ALONE VOLUMES:
Sumerian Religion (Year-2)
Necronomicon Revelations (Year-2)
Babylonian Myth & Magic (Year-2)
Book of Marduk by Nabu
Magan Magic -or- Necronomicon Spellbook I (Year-3)
Maqlu Magic -or- Necronomicon Spellbook II (Year-3)
Beyond the Ishtar Gate -or- Necro. Spellbook III (Year-3)
Enochian Magick & The Kabbalah (Year-4)
Nabu Speaks: An Alien Autobiography (Year-4)
Stargate to the Abyss (Year-4)
Arcanum Epica (a trilogy)

TRANS-HUMAN GENERATIONS

THE NEXT EVOLUTION OF A SPECIES

CHANGES

In the silence of the *woods*,
I heard the *Earth*.
In the silence of the *wastelands*,
I heard the *Wind*.
In the silence of the *mountaintops*,
I heard the *Flame*.
In the silence of the *ocean*,
I heard the *Sea*.

They all speak the same word:
CHANGE.

To *evolve* is *Life*.
To be *still* is *Death*.

EVOLVE

If the *body* stills...
If the *lungs* still...
If the *mind* stills...
If the *heart* stills...

Only *Life* is important.

Only *evolution* equals the
continuation of *Life*.

~ Joshua Free
March 2012

INTRODUCTION:
TRANS-HUMAN GENERATIONS
by Joshua Free

Dr. Timothy Leary, who served as one of the greatest pioneers of 'new consciousness' during the 20th century, said "the future of the human species is to learn how to use our brains" and that we would know the next evolutionary generation by their SMI$_2$LE.

SM: Space Migration
I$_2$: Increased Intelligence
LE: Life Extension

Although the general conceptual idea of these feats might seem 'great' in and of themselves to the average human being, one must stop and reflect if these 'parts' of the human condition are indeed where our focus and energy should be, or if it is even the responsibility and right of those from the old generation following programs of the old paradigm to decide upon. The implications of these feats has vast influence on the Next Generation (NexGen) of the human species.

The NexGen are entering the world at a critical 'turning' point, threshold or *Gate*, whereby the future of the human being will be decided...

The idea that the human species is on the 'cusp' of 'something' is almost intuitively felt by those who are alive today. It cannot always be clearly defined or explained, and nearly all of the various schools of thought, science and religion have their own interpretations and definitions whereby such a 'turning point', 'cyclic renewal' or 'End of Days' is described. One thing seems fairly certain: we are about to witness the rise of the next evolution of the human being... or its demise.

Religious teachers, social scientists and technology innovators all seem to have their own extremist perspectives on what they believe to be the absolute outcome. But, the outcome has not yet been decided – that is why we are currently only at the 'cusp' of such a transition. However, an inevitable course cannot be delayed for much longer... and that is what we are here to decide, or rather provide food for while the NexGen ultimately makes their own educated 'self-honest' decision to steer a course or fall prey to following the 'map' left to them by the older generations who perceived 'things' through an old pattern or program fixed to an old paradigm.

It is critical for the NexGen to be given the true knowledge of their origins and history so that they may see the present and future clearly...

When we consider the feats suggested by Leary and the transhumanists such as *Life Extension*, *Space Migration* or even *Increased Intelligence*, we have to take into account what the past generations using and old paradigm (program) have already handled their 'limited' handling of the less than 'meta-human' feats among themselves on the planet here. Profit-driven exploitation of the resources, human lives and home planet make the idea of *Life Extension* and *Space Migration* utterly ridiculous if the 'conditions' of life are still being handled by faulty perceptions and world-views. As far as *'intelligence'* is concerned, we are still on the brink of being able to see the world in a *self-honest* way, in which our outlook is not conditioned by a particular paradigm or fragmented 'system'.

Trans-Human Generations is not a forecast of some inevitable 'doom', nor does it follow the extreme views of those who have contributed to its ideas concerning 'right' or 'wrong'. This discourse is written for the explicit purposes of issuing reason and logic, new paradigms and observations of old patterns. It is prepared by a NexGen member of the human race who has looked forward and behind outside the paradigms of the older generation and along with the ushering in of the true-knowledge of human origins.

One can easily recognize that the human experience through time is driven by an evolutionary momentum that is constantly adaptive and rides the waves, tides and winds of inevitable change as it applies to members of the human race at a given point in its history.

What is so intriguing about *right now* is that we are on the threshold of one of a landmark transition that could involve the end of the reign of 'modern man' by some means or another – whether that be the 'post-modern' human being as a genetic extension of what is in 'power' today, or else some new form of created *artificial intelligence* that will replace us altogether and live on as the most 'human-like' being on the planet. In either instance (and these appear to be the two main courses for us to take), the days of the modern human appear to be numbered.

Without a true paradigm-shift and the accumulation of true-knowledge unadulterated by the filters and perceptions of variegated factions, the ability for NexGen to steer their own course will be thwarted as they continue to operate within the same systems and programs of the past generations who are incapable of seeing anything past their own 'fixed' parameters of understanding.

Needless to say, a population that is incapable of solving problems of living together on earth, or of using the earth without destroying it, should probably *not* be migrating into space; the masses being reduced to their slaved existence and serving as economic 'batteries' to the 'system' with little time or awareness left to their own actualization should *not* have their lives extended; and given the inaccuracy and politics of knowledge, not to mention the limited understanding the past generations have had with it, not to mention its common manipulation, does *not* really make 'hooking' the brain into an external storage and retrieval machine very appealing.

Of course, none of the originally described feats are 'bad' or 'catastrophic' for the human species in themselves, what we are concerned with is who is going to be executing these feats and how they will be dealt with. Keep in mind that it was those of the past generation that even set this course down in the first place and all of the current means of enabling them are the product of old paradigm technologies or a modification or enhancement of the same. Perhaps NexGen examination of the 'old problems' might reveal that there are more holistic approaches to the same issues that were hidden from the understanding of the past generations.

All this being said, it is important to reiterate that *Trans-Human Generations* is not a 'doomsday prophecy' for the planet. It simply acknowledges that we are on the 'cusp' of *a* change, a turning point in human evolution and its civilization, and that this course has yet to be fully decided.

Not only is the 'generational program' that humans operate on the very of cyclic renewal, but the completion of a greater cycle is also taking place: the *operating system* of the human being – meaning the genetic coding of the species that processes the programs of existence – is also on the verge of some manner of upgrade, which, if not left to external machines, should be nourished for self-actualization and evolved to handle more 'advanced' new-paradigm parameters for what is possible as *reality*.

It is and has been the current editor's hope and purpose in the literary efforts put forth under the recent guise of *Mardukite* or *Systemology* that new-paradigm wide-angle unconditioned knowledge and ancient wisdom is collected in wholeness and interpreted in self-honesty before the NexGen course is fixed prematurely.

~ Joshua Free
Spring Equionx, 2012

– 1 –
WE ARE ONE

One of the most important lessons offered from the mystic or spiritual master is the true and underlying reality of *oneness* and *wholeness* that all life and existence in the universe shares. While it may appear separated by factions, fragments and the I-not-I monad of the *Ego* on *this* level or parameter of perception, we are *all* one! We are all one consciousness; one planet; one organism; one system... and we *all* follow a cosmic pattern of beautiful intelligent design, best illustrated by the 'spiral', both cyclic and linear at once.

Many of those who contribute to the guidance and steering of human progression have done so with only self-serving intentions. Western man has yet to adopt the consideration extended by the 'Native American' philosophy of being only temporary stewards of the planet for those generations who have yet to be born. Though it defies the 'competitive' and 'profit-driven' outlook taken by the old paradigm, there is a moral responsibility here that is often overlooked – at least until it seems *too* late. Before being given their 'turn' at the wheel, NexGen will first be stuck with the 'bill' of their ancestor's actions.

14

All of the modern systems are fragments derived from ancient wisdom that have each grown from a singular tree into many branches. While they remain interconnected in genetic history, the path that each has grown takes on the appearance, albeit falsely, of being wholly separate.

The origins of human understanding, reasoning, language, civilization, traditions and systems all have an original 'trunk-line' that they stem from that is not clearly visible (nor is it going to be) from any *one* 'branch' that has emerged from it. And the longer that 'branch' continues its own accelerated expansion the further from 'Source' our perception (paradigm program) is able to see. New semantics, terminology and vocabulary to define further and further fragmentation of the wholeness are not meant to bring us any closer to 'absolute' *Truth*, nor are they intended to.

The fragmentation of knowledge and the ability to understand or recognize *Truth* is not only visible in the "realm" or 'surface world' shared by the entire human population, but it is also evident in the progression of the 'Mystery School' as it evolved in time and space across new regions and languages. These are the same 'Secret Societies' that were charged to preserve the wisdom unaltered and undefiled at their inception.

Not surprisingly, the origins of modern systems emerged from ancient *Sumer*, also given as *Babylon* or *Mesopotamia*. It is also here that we have the originating writings of wisdom, the annals of the '*gods*' initially interacting with us and the means by which they were able to raise human civilization into being. Regardless of the nature of these '*gods*' (referred to by this word not to denote 'divinity' but to distinguish from the humans themselves), it is clear in the examination of these ancient '*cuneiform*' writings and the evidence explored by 'anthropologists' that the last great 'leap' in the evolution of the human being was anything but 'natural'.

It is true that living systems evolve and adapt over time. This says nothing of 'who' or 'what' actually put the program into action originally. The expansive progression of life may very well have been put into motion by a single Supreme Being who exists outside our pre-programmed field (or parameter) of awareness. This does not discount that 'created' systems in the universe are still in 'motion' and that they follow a progressive course that we might define as 'evolutionary'. Both sides of the extreme can be easily satisfied with: God put the systems in motion and it is their natural program to continue to evolve based on that inertia (or momentum).

Given the cyclic course of human existence, we may even be able to consider the possibility that these '*gods*' could even be the survivors of an even more ancient time on the planet, one where civilization had already reached heights (apex) of technological and scientific development at least once in history and primarily disappeared from the 'face' of the Earth. For example, the '*gods*' who steered the inception of systems for 'modern man' might be survivors of an *Atlantis*-type 'lost' civilization. Others have put forth the theory that they are 'ancient aliens'.

Again, the actual 'nature' or 'origins' of our 'creators' are not the focus of this discourse so much as the general outcome and human implications are concerned. The evidence suggests that there was an evolution taking place among the pre-human species that we could denote as 'natural' and the product of adaptation and selection, but that the 'jump' that occurred to progress the evolution of 'modern man' was the product of an 'intervention'.

The human condition experienced today is not the product of natural evolution and was instead a manipulation and upgrade of a preexisting being that may or may not have ever naturally reached this point naturally in its own time.

Many facets of the human being and the human condition in a societal system can be seen to be 'unnatural' in the sense that they appear to take affect without any evolutionary cause or benefit. They are 'real' insofar as they are experienced from within the confines of the system made for (and eventually by) humans, but otherwise are not found or do not serve a purpose in 'Nature'.

Assuming an intended separation between '*gods*' and '*man*', the experience and knowledge given to man had to be 'limited' in order to keep the humans from 'outgrowing' their own place in the cosmos as seen by a differing perspective – that of the creators. Thus, as soon as the wholeness was at first collected on the tablets and in the temples of *Babylon* (post-Sumerian Mesopotamia), we find the *"Tower of Babel"* incident that shatters this wholeness forever, fracturing the 'crystal clear' wisdom of the 'Mystery School' and dividing humans against themselves with no obvious means back to their "Source".

What we now call knowledge or fragmented knowledge is really the pieces from many puzzles and no clear indicator of what the picture is supposed to look like. Humans are not educated on how *"to know"* or how to recognize *"Truth"* when it is found.

Since the "*Tower of Babel*" archetypal 'fall' of humanity, not all of the pieces of the 'grand design' puzzle have been permitted into a single system, causing each of the later emerging factions to base an entirely new puzzle (or paradigm) around one or more key elements that are links back to the ancient past. But their own puzzles are incomplete and the idea becomes: practice a generalized '*Universalism*' and expect all of the right pieces will eventually line up if we can just maintain 'political correctness' long enough to accept each and every fragmented viewpoint possible in existence as being pieces of a whole.

This is folly. For the simple 100-piece puzzle has been fragmented into several individual 1000-piece puzzles, not one of which is complete, and the acquisition of and 'dragonmind' propensity to accumulate knowledge indiscriminately has resulted in the uncovering of 8 billion pieces. The general idea has become that just a little longer spent in the effort and the accumulation of additional pieces from further fragmented viewpoints and personal opinions will somehow magically bring all the right pieces together. We become more concerned with adapting our understanding to one or several specific 'paradigms' rather than remembering that each 'system' is already a fragmentation in and of itself.

This type of understanding is important when we come to consider NexGen existence because it is currently based on old-paradigms and outdated patterns of thought and worldviews that may be 'real' and 'validated' within their own system, but say nothing concerning the holistic ideal that is necessary when choosing how to steer the evolution of an entire species. We are all one... and this *will* affect us all.

The restriction of knowledge and ancient wisdom to the 'Mystery Schools' was done to preserve the integrity of the 'systems' that were installed for raising 'human civilization' that 'modern humans' are experiencing. The 'systems' in place are, themselves, progressions of an original (or archetypal) one that evolved based on a program 'embedded' in the ancient human being that defines their own 'range' or 'parameter' of experience.

The ability to perceive or not perceive 'wholeness' or 'holistic reality' with the fragmentation of *language* (the semantic meaning of the words used to understand and communicate reality experience) was *intentional*. A limited lifespan (to 120 years), limited understanding faculties and limited movement in space were all 'safe-guards' to keep humans contained, separate from '*gods*'.

But the times are indeed changing, a turning point has arrived and we are at the cusp of becoming as *gods* ourselves. We might consider taking a lesson from them and their methods in handling *us* before we go on to spark a new race of *self-aware* 'Artificial Intelligence' that is going to replace the 'modern human being'. Keep in mind, the *gods* are recorded as first being the dominate species on the planet before the needs and means seemed to become obsolete. The time of *gods* passed to the time of *men*, but then it is also recorded that an 'eternal reign' none are given and we are to expect that the time of *men* will again pass... but what comes next?

– 2 –
STARTING OVER

When various humans speak of the coming changes, there is often little distinction made between the "*End* of the *World*" and the "*End* of the *Earth*". These are two different outcomes; both of which are possible depending upon the action taken by NexGen in the near future.

In the latter, more dismal, outcome, the Earth itself is no longer left capable of sustaining life, or at least 'human' existence. The resources critical to the survival of the human being as a 'genetic' body are thought to have dissolved or else the war-like program that drives our politics has left us with a 'nuclear winter' or another form of planetary ruin. Such could successfully end the 'reign of humanity' blatantly and outright. This is one way in which a revolving change takes place on the planet, which may even be cyclic by some other pattern that the planet follows itself. It may be that 'natural' disasters and other such phenomenon become the contributory element and that such has taken place to 'end' the progression of species in the past. Should such be the case, it is nearly impossible to predict or define what the possibilities thereafter would be.

Assuming the planet itself is not destroyed by its own programmed course in the cosmos or by the negligent use of old-paradigm behaviors continued into the future by NexGen as had been observed by past generations, then a complete social reform is just as likely to take place, whereby the alternate course is avoided by taking an opposite stance – something not together uncommon as a new generation rises up in 'revolution' against its elders as part of the self-correcting aspect of the human condition. The method or manifestation of this 'revolution' or 'reform' is not necessarily the same as what we have seen in the past, such as 'protests' and 'rebellion', but may instead be executed by 'action' and 'active involvement' in the world sought to be changed.

In the past, social reformers have traditionally separated themselves from the 'Realm' and then as 'outsiders', protest the 'establishment' in its entirety. As such, the two become antagonists to one another and much energy is spent in trying to bridge this later in attempts to communicate. But, neither side is, of course, capable of seeing from *within* the paradigm of the other and they remain mutual exclusives. There is usually some small effects from these methods, enough to still allow for the illusion of movement or progression in the human condition, but not enough.

Breaking the mold of the past generations is not the same as ignoring our ancestors. What should be kept in mind is that our recently past generations were fixed on 'dogma' to guide their own interpretation of ancient wisdom and we have seen the outcome of what their own 'perception' of reality has provided. It has put us where we are today. How much of their outdated methods of interpreting *Truth* do we wish to carry with us into a 'new world'? How will it be truly a NexGen *evolution* if we continue to adopt all of the elements from a past generation for NexGen?

The fundamental 'nature' of *cycles* ensures the element of movement and motion; it shows that certain patterns or phases are likely to or programmed to be repeated and further that the *end* of one cycle is simply the *beginning* of another. Therefore, the *"End* of the *World"* does not indicate an 'end point' of existence, but rather a reformation and restructuring of what it means to be in the '*World*', '*Realm*' or more specifically, '*human civilization*'.

Those who consider such to be automatically 'bad' or 'negative' are simply guilty of 'clinging to the old' and are resistant to releasing their hold on old paradigms that no longer serve the new and up-and-coming NexGen 'meta-humans'.

The breakdown of the 'old' is perceptibly only devastating to those of the past generations who seem to have so much of their own lives invested in the established and continued existence that they have already known. But, this says nothing of the NexGen and the possibilities and capabilities that can and should be open to them as they steer the course of human evolution at this great conjunction of potential changes.

The old programs are generally self-serving and with little regard to NexGen. They are born of a mindset that 'things are what they are', 'things have always been this way' and 'things will always be this way'. Of course, we really know that things are only what they appear to be at a given time from a particular perspective and that is what makes the generational progression so significant: each subsequent generation is given a different set of tools from which to work at a given point that none but themselves can fully understand. Each are overshadowed by leaders from a previous generation (or usually the one prior to that) and it is really the actions and paradigms of our 'grandparent' generation, far more than our immediate 'parents' generation, that generally get dealt with as reality during one's life, particularly during critical developmental years.

At the time of the inception of modern humans and the systems of their civilization in ancient *Mesopotamia*, the 'universal knowledge' that served as the launch point was not 'discovered', but 'given'. The importation of the knowledge and sciences came at once and were not the result of some development.

We see a radical change in the human condition as it moves from the primitive life of nomadic hunter-gatherers and even cave-clans into a world of language, writing, systems of socialization, cities, buildings, codes of law, commerce and class. This, again, does not seem to have followed a natural progression of 'animal evolution' and is instead the result of some kind of intervention. According to the writings of these cultures, themselves, the institutions and knowledge originally came about as a result of interaction with the '*gods*', who had originally done so for their own purposes existing outside the understanding and comprehension abilities of early humans.

Regardless of what one's opinions are concerning the actual identity and nature of these '*gods*', the beliefs installed concerning them continue to affect the paradigm of humans in their perception of reality even today.

It is important, even for the reading and proper interpretation of the current discourse, that one does not fix misappropriated definitions or meanings to the reference to ancient '*gods*', such as from within the paradigm of a particular religious or spiritual faction. For all intents and purposes, the ancient '*gods*' could very possible be 'super-humans' themselves, either left from a prior terrestrial rise to power or even time-travelers from our own future or even a distant past. It may very well be that they artificially installed a belief in ancient man of their own 'divinity' in order to maintain control, or else the limited understanding given to humans simply alluded to such a nature. Regardless, they have left their mark upon us in the form of 'coding' and 'programming' – aspects which should be examined before the human species concerns itself with charting a direction for its continuation.

The *war-program* is not evolutionary or created for 'natural survival'. It relates closely with the interactions humans had with the '*gods*' that originally upgraded humans to be 'enslaved' as a worker class, to which we may also owe the debt of the additional artificial *master-slave program* that is embedded within by necessity. Somehow, it was used to 'civilize' humans, then left to the program upgrades of language to propagate.

The *war-program* laying dormant in the human condition is a protective device that enhances the *master-slave program* used to 'domesticate' the workers. There are other far-reaching aspects of this programming that requires intense emotional encoding that also does not seem to be found elsewhere in 'Nature' and likewise does not appear to be very 'evolutionary'. Most of the aspects relate back to 'emotions', which can be experienced from one's own perception but are actually the result of 'electro-chemical' processes taking place in the human brain and body.

A functional intention for the *master-slave program* appears to be self-evident. The actual purpose is open to debate, though many scholars of ancient wisdom seem to denote that the '*gods*' were primarily concerned with the acquisition of '*gold*' on the planet and that this was the reason for a 'worker class'. Given that they were using organic bodies or 'genetic vehicles' themselves, the worker class was also taught agriculture and 'religion' so that they would bring the sustenance to the 'altars of offering' at the 'temple-homes' of the '*gods*' to maintain their lives. A paradigm of materialism, acquisition of wealth, servitude, cut-throat competition, religion and the like were artificially implanted and still go on to contribute to the human condition.

The old-paradigm motivation toward over-reproduction, profit-based systems and the blatant disregard for the environment appears to relate to human awareness of their own 'mortality'. Sadly, this tends to make the ones in power or course-deciding positions for the species more 'greedy' than those outside of that perspective. Given the tendency toward immediate gratification, these humans generally think for the 'now' and the benefits of 'today' for themselves with little regard of how it will affect the 'whole' and those who are still young and yet to be born. To them, they will be dead or nearly dead before the full ramifications of their actions are felt or even known and so they feel little responsibility for this and generally never live long enough to experience the repercussions.

When we consider the issues facing our next 'critical point', 'zero point' or 'last days' for humanity, one of the aspects that drives the species pessimistically toward a 'negative end' is the emphasis on capitalism and, of course, the distribution of perceived wealth – which is hardly wealth anymore and has since been replaced with 'paper currency' that only *represents* wealth and soon to be only 'ones-and-zeroes', meaning 'binary' or 'electronic currency'. This is but one aspect that could sure use some NexGen reform.

- 3 -
SYSTEMOLOGY IN A NUTSHELL

The wholeness all-as-one absolute existence is manifested as the physical universe. The term 'physical', as it is used here, is not meant to be considered *all* there is to existence, but simply that which lies within the range of awareness of the current human condition. Someone might argue that there are no 'absolutes', but again this is a matter of 'perception' and thereby actually proving the point originally made: that what humans call "*reality*' is based on 'subjective experience'. Reality conforms to the nature of the 'subject' and not the 'object'. Knowledge and truth, as humans understand it, thus becomes entirely based on their experience and the learned experience from their parenting generations in order to have the basis for what they consider to be their own personal experience of *reality*.

Some readers might consider these words to be a very '*inconsequential*' in regards to the everyday life they lead in this world, but it actually has *everything* to do with it. Consider that your entire range of possible knowledge and all the potential 'perceptional beliefs' within your range of experience in life as the result of *someone else*.

The term 'paradigm' composes exactly what we have described in the prior paragraph. Not only does language have influence, but also the semantics or meanings applied to the words used by the 'subject' to understand the 'object' – and by *"object"*, it is implied more than simply a tangible 'object' such as a "ball", we are referring to everything within the range of human experience. This is generally defined by societal norms of the 'Realm' in addition to general socialization among peers (within one's generation) and those from past generations.

The cyclic nature of the generations that come into being, then into power and then passing the torch to the next creates wide sweeps of movement. Between generations we can see pendulum swings toward extremist at two ends of the same paradigm. Although the 'program' of human existence is intended for 'self-correction', the tendency is to 'over-correct' in many instances; the philosophies are always shown as a dichotomy of only 'two possibilities'. Neither of the two extremes resolves the problems – which must be transformed into 'non-problems', for to deny or alter one side is to still acknowledge its existence. In this limited polar-dualism there is only *God* and *Devil* with nothing in between. How do you unlearn something once it has been given?

The human condition, being mainly unnatural in its own environment to begin with, has never reached a point of equilibrium, balance and neutrality in its existence. A paradigm of wisdom and temperance still awaits. Without such a shift in perceptions, it should not be simply technology alone that drives or sets the course for Nex-Gens. It is generally designed to accomplish the needs/problem-solving of a specific time-space in history for the people inhabiting it at that point and from within their own range of potential understanding. Two examples that we've seen drastic changes with in the 20th century past is '*transportation*' and '*communication*'.

It is a change in the human condition – a change in human understanding (field or parameter of awareness) – that is necessary in order to make better use of what is 'available', or even to expand and more clearly see the greater 'potential' of possible 'availabilities'. What good is *faster* means of transportation or communication if you are not self-honestly getting anywhere or saying anything? If the knowledge itself is fragmented, inaccurately understood, inappropriately used and even manipulated for economic, social and political purposes (all of which are misguiding), the the techno-means and speed of storage and retrieval are all irrelevant.

The *quality* of information and the ability to actually interpret it in a new-paradigm are more important than the *quantity* of how much can be stored and the *quickness* it can be retrieved. If the track that the human species is on is based on an old-paradigm methodology that leads to their own destruction (which is a fail-safe program still being accepted, previously installed into their condition by those forces and figures that intervened to *upgrade* them), than accelerating that same program will only quicken the inevitable demise.

Properly evolving the human condition for NexGen requires self-knowledge and self-honest true-knowledge of human origins and the reason 'modern man' has come into existence in the first place. From this, NexGen has a clearer ability to see and work toward the corrections that should be made toward the actualization of the 'genetic vehicle' as part of its quasi-natural ability toward evolutionary progression. Continuing to tap the same old-paradigm methods will lead NexGen into a dependency on external technologies in order to accomplish this. By not questing for inner self-actualization – which is the clear-connection between the 'genetic vehicle' body organism and the (alpha) spirit that occupies it, the true *Self* will be killed in favor of the *robot*.

The old-paradigm model currently is seeing issues, for example, with 'memory'. As such, the problem-solving ability using this model is to enhance technologies that promote improved 'memory' capabilities in humans via artificial or external means. The actual issue on *why* memory is being affected completely goes by the wayside. For example, we see a certain situation being poluted and recognize this; we then decide to take corrective chemical actions to 'de-polute' the situation, but the source of and continued element that causes the problem continues to go unresolved.

In the case of 'memory', consider that the electromagnetic and chemical causes, the acceleration of changes, exponentially accumulation of new semantics and vocabulary, the rate in which physical locales are altered, the over-stimulation and influx of bombarding messages and information, not to mention the inaccurate and irrelevant surplus of data and the inability to properly filter it, have all contributed to what humans call 'memory-problems'. None of these other influencing factors are ever resolved or even admitted from within the same paradigm – this requires someone from the 'outside' to observe (or one who is self-honest) whereas the 'logic' is lost on those already enslaved to a paradigm.

Although it manifests 'physically' as a dynamic system, the total sum of 'universals' or truth of the 'universe' are *fixed*. It is only the from the limitations of a certain paradigm of human perception and ability to understand it in parts (fragments) that is ever-changing. This movement of 'understanding' is generational, but also based on the paradigm and its 'semantic' or 'language' adopted.

Fragmented knowledge is based on a focus of 'parts'. This focus restricts the ability to see (or be aware) of *'perturbation'*, the unseen (or inability to see) the forces and factors that make things what they 'are' within the range of human awareness or field of vision. Any given paradigm immediately restricts the possible knowledge (or awareness/comprehension of it) based on a preset/fixed range/parameter/field of total possibilities (potential). This means that we have no guarantee that a NexGen new-paradigm will be perfect – in fact it is *not* likely to ever be – but if we can widen the range of what and how we are capable of handling true-knowledge, we are at least moving in the right direction. Difficulty exists in accurately predicting outcomes with a new-paradigm when we remain in old-paradigm OldGen-Tech models of understanding and recognizing the patterns before us.

It it is our ability to 'understand'; our ability to trust our own minds, senses and 'intuition'; our ability to escape old-paradigm pattern-thinking (methodology) and perspectives, that will ultimately bring about a true 'New Age' or new consciousness for NexGen evolution. Trying to update and improve on a faulty out-dated model is simply a waste of available resources and generally leads to over-corrective swings anyway.

> Principle of Generational Systemology:
> *"The most complex a system can be
> is just past the ability of a generation
> of humans to actually understand it."*

The inability for old-paradigm human faculties to see the 'wholeness-factor' of 'things' leads to fragmentation: the fractalizing of knowledge. In turn, this leads to further problems of inaccurate knowledge and misinformation.

> Principle of Fractalized Fragmentation:
> *"The greater the clarity sought of a
> fragment or part, the more unclear or
> impossible it becomes to have focus
> on the 'wholeness-factor' of the part."*

What we see in the current age is a decline, not improvement, toward 'holistic' meta-thinking.

NexGen *Systemology* is concerned with the knowledge of how things influence one another to composse a 'whole':

- How did things come to be what they are?
- How are things connected to other things to come to be what they are?
- Why do things seem to be what they are?
- Why do we see and know what we do?
- Why do we *not* see and know what we do?

The old-paradigm that guides the human condition as it stands is essentially 'reductionist' tunnel vision – not *holistic*. Humans are likely to look at one aspect in exclusion to all others, seeing it as it appears '*apparently*' in stead of seeing its connection to many aspects that compose the entire *system*. A new wide-angle paradigm emphasizes the 'whole system' – hence the term *holistic*. A fragmented understanding of *reality* being the 'closed system' that it is, gives the illusion of multiple (or infinite) 'open systems'.

<u>The Fundamentals of Systemology</u>:
- Everything is part of *The System.*
- *The System* has meta- and sub- 'systems'.
- No 'systems' exist in 'isolation.

Systems, including the 'universe' are, essentially 'recursive' or 'reoccurring', which means that from the human perspective, they manifest in cycles, which further glamours 'continuity' as being 'infinite'. At this point in history, we are seeing several 'cycles' coming to completion, as with several 'cycles' about to begin anew. This is indicative of a time of significant 'change', 're-formation' and 'evolution' and is very much akin to what many have thought of as an *End of Days*. We do not mean this as an apocalyptic message of cataclysmic doom – rather it is the *start* of a new 'day', a new 'year', a new 'age' all at once, and with it – a new *generation*.

The ultra-conservative restriction to ancient knowledge allowed for the further fragmentation of the mysteries among the 'Realm', that which takes place in the site of 'common man'. When the mysteries cannot be sufficiently kept private, they are discreetly distributed among a plethora of propaganda and misinformation in which to shield that which lies beneath. The time that has passed between has allowed for so many 'branches' to grow in their own directions from the ancient source 'trunk' that it is now nearly impossible to 'see the forest for the trees'. ...*how quaint*.

Ancient mysteries were not 'uncovered' by ancient humans, they were *'given'*. And each and every ancient culture acknowledges this. That means that apart from the 'developmental' species that led to become 'modern' man thousands of years later, there existed a separate faction of 'human' or even 'alien' species that was a 'higher mind' educated that of the lower in an attempt to educate and elevate them. Thus was born the origins of the civilization of the 'modern man' approximately 6,000 years ago in ancient *Mesopotamia*. Did something exist on earth prior to this? Yes, *of course*. But that is not what is important for understanding the current systemology of the world – and consider also, since we have not yet established a self-honest true and faithful understanding among the population of *even* the last 6,000 years, looking any further is essentially a waste of energy.

It is the quality of *data*, the means used to internally *process* this *data* and ability to actually react accordingly in the world *reality* on this function in general is in so many ways more important than the accumulation of endless *data*, the fragmented and misaligned ability to properly process *data* and how much and how quickly all of this can be accessed (such as those in favor of a purely techno-evolution for NexGen).

– 4 –
ACCELERATED GENERATIONS

There has always been a 'perceived' "gap" between the generations and while it can be ignored or embraced, the ramifications that these "gaps" lead to actually have *real* and significant effects on the 'human condition' and the systems of 'societal civilization'. The main issue that separates these *"gaps"* is the 'conscious-awareness' or 'ability to see and reason' with the given stimuli or external data received by one's environment. This is not a 'single-directional' or 'one-way' relay of energetic information, the stimuli itself coupled with the way in which that individual is able to receive the information is going to determine the 'interaction' that will take place.

So often the younger generation insists its elders '*don't understand*' them, and these elders respond with well you '*don't understand*' what it was like when *they* were growing up. Both are correct. The NexGen can never truly know what it was to grow up as a *Boomer* or *Gen-X*, whereas these prior past generations have no conception of what is involved with up-and-coming in the *New Millennium* and the far-surpassing super-accelerated existence that now plays out as the '*world*'.

When we consider the knowledge and sciences that 'jump-started' the modern systems of human civilization more than 5,000 years ago, it is from the ashes of a metaphoric phoenix that this was even possible. The planet goes through its own cyclic changes through the eras, representing a 'day', if you will. Meanwhile large social-system changes have their own cyclic periods representing, for example, the 'day'. We also have our own consciousness paradigm-shifts taking place on the 'hour', and relatively smaller or more refined cycles of socio-economical and generations for the 'minutes' and 'seconds'.

At this moment, *right now*, we are on the cusp of completion for *all of these cycles at once*, which is what prompts accelerated generation technology and the rushed dependence on either external or internal system-embrace. Everyone can *feel* that something like this is happening – but they are still unsure of their own *feelings* and *intuition* and it is just as easily manipulated, fragmented or filtered by some old-paradigm thinking. If the cycles and patterns and self-honest paradigm-shifting perceptions cannot be embraced, we will not see a 'true evolution' of humanity, but simply a radical accelerated trans-human version of an old model, with nothing to pass down to future generations but our own obsolescence.

The term 'generation' has multiple uses even in regard to the current subject matter, but unless otherwise implied, the concept being evoked in the use of this word is *'familial generations'*. The term comes from the *Latin 'generare'*, meaning, 'to beget' – and this further illustrates the most basic 'biological imperative' of the human being: to pro-create.

The *Holy Bible*, and similar *scriptures*, adopted into most Western World traditions is derived from pre-Semitic records developed and created by and for the Sumerians and Babylonians. On these *tablets*, a 'genealogy' is outlined separating the rule of various types of being, descending through the ages and marked separately by differing lifespans. We see the *gods* themselves, or else their 'avatars', descending among the physical world for *tens of thousands of years*, followed by demigods for *thousands* of years and the earth born deities for *hundreds* of years. By the time of 'modern man', the 'human life-program' was intentional restricted (genetically) to 120 years of operation, a number significant to the original sexagesmial mathematics of *Mesopotamia*.[*] The tablets indicate that the quicker a reproductive age was met, the shorter the human life would

---

[*] *Mesopotamia* – referring to both *Sumerians* and *Babylonians*.

be. This indicates that the programmed accelerated changes endured during puberty actually contribute the low lifespan of the human being.

Modern terminology calls those who live to see 100 years in life, *centarians*; and those who live to see 110 years, *super-centarians*. Only one recorded life has ever exceeded the 120 year 'biblical' limitation: *Jeanne Louis Calment*, who lived 122 years and 164 days, from February 21, 1875 to August 4, 1997. It is also of considerable note that the vast majority of 'longest living' people in history are female. The two factors that contribute to the duration of one's life-span program is the environmental and external 'stress' factor absorbed from the world and its changes in addition to one's own genetic program, which the body runs on, including its own immunity and other signatures. The 'alpha-factor', which relates the the 'spiritual identity' that resides in the 'body', using it as a 'genetic vehicle', is also a factor.

Naturally, there are external conditions, such as quality of life, air purity, lack of chemical additives and pollution, water purity, ceasing to experiment with biological warfare and such that would also either prolong or detract away from a given body's ability to 'survive'.

The NexGen are coming-of-age at a time when old-paradigm science and its scientists have only just realized that they have been stumbling upon one another, each straining in its own direction in its own 'system'-restrictive way of perceiving the world. The modernization, industrialization and otherwise *'Westernization'* of the planet has resulted in the rise and fall of perceived economy and the roles of modern generations in cyclic patterns – a cycle that is going to continue on indefinitely to seeming *nowhere* if not recognized for what it is, and *changed*.

Throughout modernized history, the average age of a female at their 'first birth' (16-28) is 22. Other records have indicated 20. We can easily settle at 21. This means that *every 21* years, we see a *new* generation born and a generational shift in social 'roles' by all those existing generations. A 'new-born' generation comes of age and become 'young adults' at 21, whereby a new 'new-born' generation enters the stage. Those who were 'young adults' reach an age of 'mid-life' or 'maturity', and the 'mid-lifers' become 'elderly' and so forth. Socio-political, world events and technological and consciousness cycles all contribute to specific 'periods' whereby one can determine where their own lifespans have fallen among the generational distinctions.

<u>1946-1964</u> "Baby-Boomers" or *Boomers*
<u>1965-1981</u> "Generation-X" or *Gen-13ers*
<u>1982-2000</u> "Generation-Y" or *Millennials*
<u>2001-2020</u> "Gen-Z", "NexGen" or *Gen-infotech*

This being said, there are those who are born along certain cusps (64-65, 81-82, 00-01) that are going to '*feel*' or retain aspects from both sides of the 'generational line' distinction. The acceleration of technology and the progression of the information-age has also widened the perceived 'generation gaps' between the distinction lines. What is available or within the awareness potentiality for the *Millennials* and its development during their own growth-years is far accelerated from what was present for *Generation-X*, which is exponentially accelerated beyond what was available to the *Boomers*.[*]

Although considered arbitrary pursuits in the past, the trend observed after decades of generational studies now shows that socio-economic and political reform coincide with the rise and fall of generations and that both are cyclic.

---

[*] The *Boomers*, now 48 to 66 years old in 2012, are the extent in historical generations for which the current editor intends to include in the discourse, which is, for all intents and purposes geared toward heightening the awareness of the younger generations, such as the *Millennials* and *NexGen*, ranging from 30 years old to not even-yet-born.

Generational 'gaps' or *distinctions* are not only divided based on the plotting of dates on a calendar, but also the general 'social' outlook 'worldview' and 'perceptual range' (conscious-awareness), which contributes to the cultural trends, tendencies and other social divisions. Consider the tendency and mentality to 'settle down' or get married at a young age versus a worldview that prefers to "date and rate" many potential suitors and resulting into various 'alternative' lifestyles (by old-paradigm standards) that extend into adulthood. We might also witness high 'religious' ideals transform into and emphasis of high 'spiritual' and 'gnostic' patterns.

New economic patterns, which were actually cyclic from previous periods of techno-industrial revolution, forced *Generation-X* and early *Millennials* into a period where the social-system came to raise them; not parents – or else they were left to mainly raise themselves. Different parenting 'schools' of thought came into being, but this mainly went by the wayside as more and more 'latchkey' children walked their developmental years, with both parents being forced to maintain one or even two jobs to make-up for not only the economic *'inflation'* of the dollar, but also the rise in technological needs and its corresponding *rising* costs.

The '*digital division*' separates the previous generations from the *Millennials* in several respects. Firstly, there is what is considered 'widely available' during developmental years. The kind of technology now accessible to the average child was either only available to the upper echelon of society in the 1980's or else unavailable altogether. Even putting the general 'computing' power aside, the *cellular phone* is a perfect example. Considered a symbol of status or importance, the non-military use of *cellular* technology in the 1970's and 1980's was most restricted to upper class businessmen. The *cellular pager* was originally a stereotype of medical professionals. These technologies later became commonplace in a way that is so ingrained into consciousness that we have entered a time where we cannot place ourselves (as up-and-coming generations) outside of a time when these technologies were not so available.

Modeled from "*Star Trek*" communicators, *cellular* technology not only reveals its physical or technological nature by its name but also what it attempts to connect among the population as a microcosm – linking the cells of the human race. In this way, each human carrying one becomes dependent on it for their interconnection to any and all other 'cell' entities on the planet.

The NexGen tendency is generally observed as leading away from a centralized home phone and to connect identity to 'cellular' or even "e-mail" as their primary communication 'addresses'. This has created 'new' technologically involved avenues of tech-specific or specialized or limited economical growth, while at the same time subtracting that which drove other affected areas of commerce – but this is not the area of which the current topic is to dip. The illustrative point being: what is familiar, commonplace and within the field of awareness to one generation is not the same as another and can only be seen from and within that generation as being 'truth' from their perspective.

The *Millennials* and few of the *Generation-X* in the population were not alive to see the initial social costs of 'research and development', nor the acceleration of technological pursuits for its own time relative to the even more rapid development and speed for this age. Though even now we are seeing only a progression of the same old-paradigm technological pursuits, they simply are getting *smaller* or *bigger*, *flatter* and more *fashionable*, holding *more* and accessing *faster*... but essentially doing all the same things they did for us in the 1980's and 1990's – especially after "*America OnLine*" launched in 1993.

Not surprisingly, at a critical cusp of a sub-generation that divides the *Millennials* in two parts, the 'internet' has been publicly served since CompuServe's first public e-mail services were offered widely in 1990. Increased accessibility, in addition to other dial-up modem offerings were increased with AOL in 1993, including direct access to elitist 'usenet' groups that had begun forming in the 1980's. Where once customers were charged by the minute for their internet service, new unlimited and accelerated methods have since developed, including a rise in 'hot spots', where access to *wireless internet* or *wifi* have become prevalent. So, even the speed of having hardware capable, to the formation of a static internet, to the now constantly available '*always on*' portable or remotely accessed ever-changing *Web 2.0* is becoming the 'norm'.

Aside from the 'newgroups' and 'bulletin boards' that had always been in use, the original internet was similar to a very large book, wherein the pages may be added or change slightly from time to time, link to other pages, receive small updates and the like – it was mostly static. People looked up specific information or checked out what local groups and organizations had posted concerning their success, but it was not yet a 'way of life' – it inevitably would be.

While it is now taught to grade-school children as a supplemental to their other more traditional educational pursuits, the means of web-page construction was at one time, and remains to be among the past generations, a 'specialized' knowledge that many profit(ed) from. Further, as illustrated in the 'rise of the .com', the purposes supporting the existence and function of the internet have always been *commercial* – and given the media being employed, these are considered *advertising* or *marketing* oriented. Either we see 'ads' themselves or else a direct means to purchase items. We have witnessed economic pros and cons to this and continue to do so today.

The realm of *e*-Commerce and the rise of the interactive "Web 2.0" lifestyle of social networking and blogging has all contributed to an already growing tendency toward 'immediate gratification' and the ability to manipulate environment (or 'augmented reality') in ways that fixate on the stimulation of pleasure-and-reward centers in the brain... such as we see with video-games or even many of the drugs in 'recreational' use. Impulse-purchasing and the circulation of money for what would otherwise be considered unnecessary or outside the realm of accessibility seems to profit the avenues and sites (like *e*Bay or *PayPal*) far more than average people.

As has been seen in the past, the NexGen movement of conscious-awareness and social systems reform will be guided by the different experiences that they have based on a different means, accessibility, the social leadership in power at that time, not to mention the beliefs and self-constructed barriers or 'baggage' adopted from the previous generations.

This type of social issue has been studied by social psychologists and economists for decades and has been clearly illustrated by the work conducted by the two men who actually coined the term *"Millennial"* for the recently up-and-coming NexGen in the books *"Generations"* (1991) and *"Fourth Turning"* (1997) by William Strauss and Neil Howe.

– 5 –
EVER ONWARD:
GENERATIONAL CYCLES

Generational '*gaps*' or '*distinctions*' are related to an age-group of the population as it reaches critical points of its own lifespan phases in combination with specific points in history. A person who is 21 years old in 1969 is experiencing a radically different *reality* than someone coming-of-age into the *New Millennium*. Never the two shall meet and all of the 'grown-up', 'adult' and worldly experience the *Boomer* has to offer NexGen is based on their own paradigm and worldview, specific to developing (or growing through) a time in history that is no longer existent.

This distinction or separation being made clear, this is not to say that the parenting generations currently in adulthood or elderly phases of life have no 'experiential wisdom' to offer, but the truth of the matter is that *times have changed* and with this must change the paradigm of thought. For although the universal knowledge remains at a constant, we are experiencing a glass (unseen) ceiling to what the old-paradigm methods of approaching *life* and *reality* can earn us. We move onward seeking motion indiscriminately.

A complete generational cycle, according to the most clearly illustrated models, is four-fold. It defines the sequential generations as a cycle that begins anew with every four – which means that approximately every 8y4 to 90 years, we are seeing a new generational paradigm-shift. We are currently also facing a 'New Age' consciousness shift that is recursively trying to raise the awareness of 'modern man' to 'ancient wisdom', such as has been lost with the 'Dark Ages' or even prior with the destruction of some of the oldest libraries that once existed.

In addition to a four-fold generational cycle, the manifestation on social consciousness results in equally four different phases of a socio-political and economic cycle. The two cycles compliment one another as well. Before relaying the distinctions made by social scientists as to where the generations we are most concerned with lie, let us first give a brief examination of what the four generational archetypes and four phases of modern society actually are.[*]

---

[*] An understanding that has been primarily been refined by the late William Strauss and his associate/co-writer, Neil Howe, authors of *"Generations"* and *"Fourth Turning"* with the corresponding LifeCourse Associates business/organization.

1st Quarter: The High
2nd Quarter: The Awakening
3rd Quarter: The Unraveling
4th Quarter: The Crisis

Just as a human lifespan, or more appropriately, the cycles of nature in the seasons, this could be considered the same pattern we see with: new growth in spring; maturity in summer; elderness in autumn; and recursion or reset/hibernation in winter. It is generally considered that we are currently on the cusp of a *Fourth Quarter* Crisis for our socio-political and economic condition in society, something we would be expecting to see 'four quarters' ago – which we do, at the last Great Depression of the early 1900's.

<u>The High</u> – The cycle begins anew following a period of crises, such as that which came after the last period of *World Wars* and produced the *Boomer* generation from 1946 to 1964. The theme that drives this period is marked by strong institutionalism, weak individualism and a confidence in a general 'collective' social norm.

<u>The Awakening</u> – A phase of the cycle marked by extreme over-correction from the previous era of guarded conservatism, such as what we see developing from the 'consciousness' movement

of the mid-1960's until the early 1980's, marked by the *Generation-X* and similar generations at this point of the cycle that uphold personal autonomy and self-reliance where the up-and-coming adults reject the adult world and provide enthusiasm for social reform. *Boomers* become '*Hippies*' who given birth to '*Generation-X*' and other '*Millennials'* further down the road.

The Unraveling – A period of 'fruition' whereby the up-and-coming and adult generations can enjoy the result of the 'Awakening'. This period is considered the 'beneficent' and relatively prosperous period of 'peace' between the energetic reformation of the Awakening and what eventually results into the Crisis. This would be the equivalency to the technological progression and economical heights experienced from the mid-1980's through the 1990's until the period of Crisis fell upon us in the *New Millennium*.

The Crisis – A period of death and eventual renewal of societal systems, secular upheaval, the replacement of an 'Old World' or 'Old Deal' with a 'New Order', leading to reform via revolution, civil war, world wars and eventually a socio-economic '*Great Depression*'. Period begins with a threat of national survival indivative of 9-11 in 2001 and economic collapse in 2008.

Each of these phases of socio-political and economic history *beget* a 'generation' of a particular archetype. As such, each of these archetypes age and progress in their own life-cycles as new periods or phases of the world take place. The four generational archetypes are essentially:

1. Prophet/Idealist: born during a High
2. Nomad/Reactive: born during Awakening
3. Hero/Civic: born during Unraveling
4. Artist/Adaptive: born during a Crisis

As such it is *Boomers* who appear as the idealists born during the post-War *High*. They enter a world denoted: "America: The Superpower" and usher in the time of *Generation-X* who are the in-between or transitional 'nomads' born during an 'Awakening' marked by the last great 'Consciousness Revolution' (denoted for its pop-culture 'New Age'-ism). The *Millennials* are the 'heroes' born from a period of Unraveling, which has been experienced during the mid-1980's until the *New Millennium* as the 'Culture Wars': the development of the same sub-cultures and counter-culture that the *NexGen* adaptive/artists are wanting to see broken-down to make way for a globalized-unity and technological *Singularity*. All of this has been quantitatively predicted and foreseen – but, *what* do we do about it?

Following this generational cycle, the idealists give way/birth (*beget*) a generation of realists, who in turn give way/birth (*beget*) a generation of innovators, which follows by a generation of adaptive artists. The cycle begins again with the artists giving way to a generation of new idealists of a 'New Age' – and so on and so on.

Those who are born during certain periods, or rather, come-of-age (up-and-coming 'into') during certain periods are going to 'absorb' or 'receive' lessons and critical aspects of human social and worldly development that are specific to those periods, but which later will carry forward into later worldviews, behaviors, beliefs and atttudes in their own lives during other phases. Each of the generations, as much as they might like to mold another in its own image, actually begets an 'opposing' archetype. The NexGen of any given generation will always work to correct or compensate for what was perceived to be the problems, excesses or inadequacies of the midlife generation that is in power. For example, even though it is the *Millennials* who are up-and-coming into young adulthood, it is the *Boomers* who are still in position of authoritative power, leaving *Generation-X* in between seeking to find and maintain its own footing while giving birth to the *NexGen* who are coming into being.

The current state that we are facing is the apex of the 4th Quarter or Crisis period whereby the societal and economic systems that were born, matured and enjoyed in fruition over the past 80 years (since the last great economic collapse) are returning; we are living in it now and it is still going to be playing out its process for some time to come. It is not an all-at-once 'event' such as many people think of when societal changes take place, rather, it is an evolving change that progresses over the course of a period of time and is often not even noticed for what it is until it is 'too late' or so 'apparent' that it is interpreted as a 'problem', or more correctly, the 'crisis' that it is.

The term 'crisis' is generally given a negative connotation and this, of course, is subjective, going back to the age old argument of what is good or bad is subjectively based on who or what is being helped or harmed via some 'event'. From the *Greek* language, '*krisis*' actually means 'critical', denoting a 'critical point' that is being reached that will have implications toward what will happen next far more than an 'uncritical' point. They are generally considered negative, not because of actual 'pain' or 'detrimental' to being alive, but because they radically or 'critically' change the way something is being handled.

Right now, the human condition is be challenged by its own 'crisis', a critical point where we are wanting to 'break down' or 'reform' what it means to even *be* "human" for the next generation. Likewise, we are seeing a critical period of change or 'crisis' on other system-based social levels such as the 'security' of our privacy as individuals, the economic or monetary systems that the others are based on, the way in which socialization and politics are being looked at in addition to the obvious effects that human 'trial and error' throughout past generations have had on the physical environment of the earth – the very planetary home for *NexGen*!

More than simply stating that we are at a cycle end, being 'blah' about it and going with the flow of the minority that are in the control to make decisions on behalf of the greater population, many are choosing to look at this critical time as a 'positive' thing; not 'apocalyptic'. A deeper look at the etymology – or word origins – of crisis reveals that it is not nearly a time of disaster, but a 'critical' time for making changes about the future: "a separation, power of distinguishing, decision and choice, forming a new judgment: to pick out, be selective, choose, decide, judge and formulate" a new path. Now is the time to chart the new path for humanity!

59

A 'crisis' or 'critical turning point' as related to Systemology is seldom considered at points when a person or social collective has reached this 'state'. The 'crisis' will usually only be found in 'complex systems', such as the personality-persona-program that humans consider their own 'identity' or else with broad 'social systems' that are used to support the human condition: economics, politics, social-order, etc. It is a point where the 'system' has been shown to be functioning 'poorly' and/or with truly devastating effects, but which, for one reason or another, will not 'break down' or 'collapse' naturally, it just continually 'disintegrates'. When the 'crisis' is acknowledged, suddenly it is a 'problem needing to be dealt with'. No doubt the issue was present for a long time, or even since its inception, but the system continues to be allowed its existence until it is wholly 'apparent' that there is 'something wrong'.

Rather than attempting to avoid the inevitable and be ultra-positivist in the approach, the current author/editor has always joked that "am here for 'damage control' in these times being what they are". The idea of avoiding the crucial point in entirely unnecessary and a waste of energy. Spending time/energy this way causes depression because we don't *get* what we *want* by play 'ostrich', burying our head in the sands and looking the other way thinking, "well, if I don't see it, acknowledge it or believe it then it can't possible affect me". Really? How well has that worked in the past for real? Can we seriously expect the entire planetary system to just 'ignore' us as we continue selfishly and oblivious to the 'bigger picture'?

In an article prepared for 'crisis management' on an economical/marketing level for businesses, the famous quote emerging from S.J. Vinette in the article on "Risk Communication" (2003) clearly defines this point of 'systems crisis' as: "a process of transformation where the old system can no longer be maintained." But more than anything else, the defining factor of the term or 'turning' in human history, evolution and planetary life seems to signal that clear and present word: CHANGE.

CRISIS = CHANGE

We are told that we need to 'keep' moving forward as a species in order to survive – but movement has many forms and it seems to be that only the most highly external, obvious and worldly and profitable means are being explored using old-paradigm methods on the NexGen 'seekers'.

The modern and post-modern NexGen are given a 'bill' to pay based on outdated standards of even what this very statement means. In one, we think of the 'financial bill' that comes as a particular cost for this or that. On another level, the laws and social codes mandated by the '*Realm*' are often introduced under the semantic title: a bill. In either case, this 'passing the buck' forces a later generation to assume the responsibility and ramifications for what a twice removed[*] prior one has done. The inability for a NexGen (even those who have been found in similar circumstances in history past) to succumb to the paradigm given by an past generation results in what many people call: *revolution* – which is just as 'scary' of a word to some as '*crisis*'.

Both the '*crisis*' and '*revolution*' are based on ancient words to indicate a 'sudden change'. This can be 'scary' for those highly dependent to an existing systemology or those unwilling or unable (due to their 'current' faculties) to change.

---

[*] Meaning that, for example, the *Millennials* will end up dealing with the brunt of the decisions made, not by their immediately former (*Gen-X*), but by the *Boomers*, who are currently the primary generation in power and with direct mass decision-making abilities. The *NexGen* proper, will then be dealing with the choices made or ignored by the *Generation-X* as they begin to assume more authoritarian social roles (or at least by at an age to be competitive).

The fallacy generally adopted in pop-culture, thanks to the efforts of the past few decades outright 'visible' or 'apparent' progression is that: *all change is good.* Any logical person should be able to discern that this is indeed a fallacy – a logic leading to faulty reasoning, and for the purpose of the continuation of the human species, could be outright detrimental. But the pressure to 'save ourselves' by making '*a* decision' comes and requires execution in a timely matter if an unwanted outcome is to be detoured around.

When we refer to 'revolutions' in the past and future, it is important to understand that this is not always indicative of the kind of 'outward', 'visible' and 'apparent' *"riot in the streets"* type of 'revolution' or 'reformation'. Though from a similar word source, what people are really afraid of during these eras – and what is being described prior in this paragraph – is 'revolt' or 'rebellion' that incites a 'visible', 'physical' *"social uprising"* among the population, which can even turn a country on itself in "civil war". This is not the most productive manifestation, regardless of the success or reasoning that it may have held in the past. If America succumbs to 'civil war' and 'brother-against-brother' uprisings with 'people against the state' in rebellion we most likely will be destroyed by an outside force in the process.

– 6 –
TRANSHUMANISM:
PROGRESSION vs. EVOLUTION

The current generations are facing a new development that for a long while had only existed in the science-fiction fantasies of the older generations. But one man's fantasy becomes another's religion (and *visa-versa*). The 'idea' of transhumanism is as old as the 'modern human being' in general, but it seems that it was only with the paradigm possible through the last century or so of human thinking when humans, or at least the more 'enlightened' or those with increased 'awareness-consciousness' were beginning to be able to see the new possibilities or potentials for the human condition. Some of these capabilities are already upon us and to their credit, they are the product of generations inspired by the technological 'potential' of human innovation. But, rather than 'throw the baby out with the bathwater', it is a cry for 'balance', 'holism' and 'tempered wisdom' that this discourse should echo, not simply a xenophobic fear of all things technological.

Humans need to be given the skills and resources to actually see and think... what a concept!

"Technological Transhumanism" (often represented among such thinkers as **h+** or **H+**) is a posthuman movement rising in culture and consciousness even as you read these words. It applies the same ultra-progressive accelerated expansion of 'technological capabilities' to the human condition, but in the most direct, outward/external and 'apparent' ways. Rather than actually enhancing the 'genetic being' and 'genetic memory' that is passed down as an evolutionary change for NexGen to be biologically transhuman, the common tendency is always to supplement, correct or enhance the body by an "artificial" means.

The type of accelerated technological advancement that drives a purely "Techno-transhuman evolution" seeks to enhance or change the human condition only as it 'seems' or is 'apparent', thereby leaving little development for the species outside of the current paradigm. Becoming entirely dependent on artificial means to experience a NexGen 'augmented reality' (or 'virtual reality') still relies on the systems being in place that fuel and maintain such technologies. It says nothing of our survival compared to the power of 'solar flares' or if the 'grid' should collapse. All the development efforts are restricted in the 'Realm' to external, profit-driven, 'concepts'.

We are on the cusp of reforming or establishing a new 'baseline' for what it *means* to be 'human', and that is what 'post-modern', 'post-human', 'trans-human' or 'meta-human' thinking has always been about. Futurism once meant that we would shy away from 'crude dependencies' and usher in a period of peaceful near-Utopian standards, but the 'human element' keeps getting in the way of that. The *master-slave* programs that fuel the greed and emotional instability toward irrational ends keeps 'modern man' in its place; and if not to some *god*, then to some *man*, but always a perceived 'authority' left in power to dictate the 'reality' of the 'realm'.

Although transhumanism of the 1960's, as described best by Timothy Leary (at the opening of this book) with the SMILE toward *space migration* and *increased intelligence* and *life extension*, it is perhaps the old-paradigm way of interpreting the message that keeps us going back to the now familiar and comfortable 'pseudo-security' felt with external technologies. A direct movement toward these predefined 'ends' for a new progression (program) of the human condition could just as easily begin with an emphasis on self-actualization, something more akin to what Nietzsche was describing as the *ubermensch* or '*overman*', the true NexGen 'genetic' evolution.

It was actually another of these super-consciousness pioneers in the 20th century that actually coined the term, *transhumanism* – none other than Aldous Huxley (author of *Brave New World* and *Doors of Perception*), who defines this concept originally as: "man remaining man, but transcending himself by realizing 'new' possibilities of and for his human nature". And Timothy Leary's original codex for this progression was that the "future of the human species is to learn how to use our brains", but not necessarily rely ever on a 'machine' to '*use*' it for us.

The simple fact remains that inevitably the human species will *beget* a new 'evolutionary' post-human or trans-human generation, by definition though. This does not guarantee that the more intelligent species will be a direct 'offspring' of the human being, but perhaps a genetic or technological 'creation' or 'intervention'. At first, the new 'breed' would be brought into being in order to assist (supplement) the current human condition or population. Eventually it would supersede it and render the human being obsolete. This would happen regardless, assuming the next 'genetic evolution' of humanity, such as we have seen in previous models. 'Modern Humans' would not necessarily *evolve*, they would simply step 'out of the way' and go *extinct*.

"Technological Transhumanism" and "Genetic Transhumanism" appear to be two different schools of thought, each locked within their own systematic paradigm and each fighting one another for control of human destiny. They do not necessarily *have to be* two mutually 'exclusive' options, but given the old-paradigm self-correction tendencies, it seems like we live in an all-or-none 'realm' where, as with many things, it must be *one* or the *other* and never the two shall meet. Where the first seeks to enhance human life by interfacing the mind/brain with technology as a supplement or 'tool', the second one is concerned with the 'human being' as a 'genetic vehicle' and the types of traits, progressions and abilities that can be enhanced and passed down genetically concerning the mind/brain and one's understanding of its use.

An internal method would involve a person having a 'higher' understanding of how their brain works and the abilities of the 'mind' in relation to their conscious-awareness, improved critical thinking or reasoning skills and a refinement of perceptual and emotion-based reactive thinking. Senses could be heightened. Inclinations and intuition would be clearer. Personal barriers and self-restrictions would be dealt with in a way that still improves the quality of *all* life.

A purely external mode of thinking and causal action depends on the use of technology to eliminate 'human' suffering (with usually little regard for '*all* life'); life extension occurs using technological means of 'immortality'; the ability to upgrade or self-modify to the point where gender and the family unit become no longer viable coupled with the establishment of super-intelligent 'Singularity'.

In either case, we could see an era of 'new' (or very 'old') religious or spiritual movements and efforts to become or awaken the 'dormant' and almost 'alien' "Self" that resides within the body to use it as a 'genetic vehicle' for its own advancement. The philosophical program of the brain-mind has never fully satisfied human understanding and thus the next evolution or NexGen or progression rate still seems to point us toward a fate that is 'more than human'.

One of the foremost social lecturers and cultural figures involved with technological transhumanism today is "Max More" (born Max O'Connor), who stated: *"No more gods, no more faith, no more timid holding back. Let us blast out of our old forms, our ignorance, our weakness and our mortality. The future belongs to posthumanity."* But, just what kind of post-human future will this be?

The post-human 'technological' era is generally known as the "Singularity" by those who pursue this philosophy (or systemological semantic vocabulary). Although it denies the Self its own evolution by relying solely on an external means of 'personal development' (which is only temporary, and not genetic), is it even avoidable assuming we continue the current paradigm? Will humans simply do it because they can, regardless of what the consequences might be? It seems likely at the accelerated rates we have been seeing technology moving in just the past few decades. We can only imagine what might be readily available in even another decade or less.

The kind of 'intervening' progression that will be witnessed by NexGen is not necessarily 'natural' or 'evolutionary', but simply the following of a program preset over 5,000 years ago whereby humans would be limited from reaching a 'higher' state than they were. Sure, the human is fashioned in the 'image' and 'likeness' of its last 'intervening' force (or '*god*'), but it was left with enough fixed parameters to keep it from being a nuisance during its evolutionary progression, in regards to other life in the cosmos. Metaphorically, to keep the from being able to readily fly away or to the abode of the 'hidden *gods*', their consciousness wings were 'clipped'.

The kind of 'intervening' influence that is likely to assist in the accelerated 'evolution' or 'transition' to a post- or trans-human *Singularity* is something that we find evidence for when humans were last upgraded with the programming for modern 'civilization' in c. 4000 B.C. What is taking place, just as in the anthropological missing link found much earlier to this when humand were only bred for 'physical work', is an intervention that is far faster and more accelerated than either adaptation or natural selection. Both are valid when things are allowed to assume their normal pace, but that is not what is the case for the human condition, which is in itself an 'intentionally' accelerated species.

This '*Singularity*', should it transpire, would be marked by the human species being able to create a 'near-perfect' (subjectively speaking) machine that surpasses all known human ability or 'field' of conscious-awareness. This ultra-, super-, meta- *intelligent* machine makes all human-based ideas and innovations obsolete under the 'new-paradigm' or 'futurism' under the 'vision' of an artificial creation. Further innovations and 'invention' would come specifically from the super-intelligent machine directly and would be considered beyond the 'control' or 'rule' of humans, for which it will 'choose' *not* to serve.

By old-paradigm standards, humans, with their sensitive *Ego* and 'bullish' pride will have grave difficulty in accepting even the possibility that 'mortal humans' could give *birth* to 'machine sapience', even though we have been 'gradually' escalating our resources, knowledge and technologies toward such a 'pinnacle'. It is already happening around us now with 'lesser' 'old world' technologies that have already automated enough tasks to feel the lack of human worth in many instances already. This leads to an economic collapse created by the masses who are no longer able to viably 'function' in the societal norm or 'New Realm'.

The effort that will push it over from 'seeming harmlessness' will be a 'final *tweak*' of lesser technologies already in place that will somehow be able to all come 'alive' or 'online' in Singularity in a moment. Theoretically and simultaneously, everything that the 'system' now relies on to function would be automated enough to keep it from being 'disabled'. A truly external trans-human technology could not be controlled by a 'lesser mind'. Any rules, such as those put forth by Isaac Asimov's *iRobot*, are purely idealistic and limiting compared to what humans of the old-paradigm would prefer: self-correcting with limitless potential... *just like we were made...*

Without focusing on or balancing the factors important to the 'genetic being', the evolution of the mind (conscious-awareness) and its abilities that can be encoded in the 'genetic structure' of the individual – something that can be passed down directly to subsequent post human Nex-Gen – we have only two likely outcomes resulting from the current old-paradigm untempered drive toward 'total technological transhuman evolution' is 'Human Extinction' or 'Human Enslavement', neither which seem to promote an increase in the quality of the human condition.

'Human Extinction' results when there are no longer enough biological entities carrying this genetic structure that we call 'human' today to continue their species forward. This scenario appears most likely if the new technology, transhuman 'life-form', or 'robot' is 'human-like', thereby becoming the next 'human-like' evolution on the planet. By the time of our best creations possible, being able to surpass this will be impossible and all new 'innovation' and 'invention' would be the product of our creation. 'Human Enslavement' is less interesting. Being seen as obstacle leaders on Earth, they would be replaced with 'trans-human machines' that enslave the existing 'humans' by force, usually for the purpose of 'early-stage production labor'.

– 7 –
NEW WORLD – NEW DAWN

The NexGen faces being born into a world of uncertainty from where we currently are standing. Nothing has, as *yet*, been decided for our future beyond all doubt. There are many patterns, cycles and trends that do seem to suggest possibilities more logical or likely to occur than others, depending on what course is taken; and when.

An uncertain future is not a bleak one. And even though 'change' can be stressful, it is not necessarily a 'good' or 'bad' thing except that it promotes the increased quality of all *life* on the planet. On the one hand, we could play 'devil's advocate' and say, well, humans have messed up the earth long enough, so let them create their own demise by allowing a sentient technology take over the planet. However, there is no logical reason to believe that an 'artificial life-form' is going to be any more ecologically or environmentally 'conscious' as old-paradigm humans.

One of the reoccurring prophecies concerning the reform of a 'New Age' following the collapse of the 'traditional systems' all surround our attitude of money and currency and what it brings.

Assuming a complete reform of our world, the entire culture-base, political-base and economic-base would be in '*crisis*' to the point where the only satisfactory solution would be a new-paradigm; a new approach; a new vision; a true 'New Age' NexGen perspective on what can be done. The old patterns don't provide a new paradigm. It is almost as if it requires a 'start from scratch' by the time of this cycles end/beginning given its rapid acceleration in the past few centuries and the decades toward a 'zero-point' where the 'old ways' can bring us no further as a genetic species.

The kind of techno-industrial specialization that we have seen since the mechanical revolutions in the last few hundred years breed obsolescence by nature. Humans are forced into lives of complacency, too 'bogged-down' by their work-a-day life, feeling stalemate and idle with the long-term futuristic ideals once held now lost in the constantly bombarding details and noise of the current 'information age'.

Without the importance and worth of 'Self' and sense of individuality and creativity being lost to the 'Grand Machine', new meaning and new purpose must be 'brought' (not 'sought') to those in 'positions' to make/manifest the difference.

Most importantly, it is the ability to maintain a sense of 'centeredness' or 'peace'; our ability to actually *cope* with the times being what they are manage ourselves, being responsible as an individual *and* as a collective, which will determine the outcome. Essentially, nothing is guaranteed.

These feelings of 'mystery' and 'uncertainty' are simply the 'unknown' coming ever present, closer to an inevitable 'launch point', whereby we will be able to shed skin like a snake, freeing the *Self* to be the beautiful meta-human identity that it truly is. As some 'climax' is waited for or even sought, we see a tendency for those to drift away from the systemologies and paradigms that are no longer serving them or have been found to be faulty.

There is increased attention being given to the 'higher orders' of learning accessible to humans, but this seems to come from the 'underground', meaning outside the surface world of lights and media. It uses the same means and media now, including technology and the internet to relay its message – but this is demographic oriented too. Given the high volume of *Gen-X* and *Millennials* occupying the internet, the messages intended specifically for them are likely to use that media, such as the "Mardukite" movement.

It is important for *Gen-X* (age 30 to 50) and Nex-Gen, including the *Millennials* (age 10 to 30) to work together during this next period of *crisis* and *upheaval*. While the *Millennials* rise to power positions of a new version of what we have or become heroes of reform, the communication between generations is important to keep the entire human existence from falling short of its own potential in the 'next' cycles.

We are already seeing signs of an advanced or accelerated type of human being born even now. Some distinguish them with fancy 'New Age' names like 'indigo children', 'rainbow children', 'star seeds' and the like. Although it is not the current subject being discussed, it is a microcosm of the worldly issues being addressed and an illustration as to what *could* be possible for the human evolutionary future should we wish to put more of our focus, attention and resources towards the personal and genetic progress that can be made with simply the 'human being' itself. Humans are easily forgetful of where they came from, what got them to where they are, and that things as they are or seem are not all there is or was or will be. As NexGens continuously rise with higher faculties to understand, we may just get to recognize them by their SMILE yet.

WOULD YOU LIKE TO KNOW MORE
? ? ?

Discover other great titles in the
"Systemology Series"
by Joshua Free

including

"Human, More Than Human" (S1)

"Systemology: Defragmentation" (S2)

"Alien Agendas" (S1+S2 Anthology)

"The Hybrids – A Novel"

"Nabu Speaks! An Alien Autobiography"

...and don't miss the original underground
classic "Mardukite" literary legacy
also by Joshua Free.

www.ingramcontent.com/pod-product-compliance
Lightning Source LLC
Chambersburg PA
CBHW051221170526
45166CB00005B/1993